Graphing
GRADE 5

Written by
Robyn Silbey

Illustrated by
Corbin Hillam

Cover Illustration
by Susan Cumbow

FS112039 Graphing Grade 5
All rights reserved—Printed in the U.S.A.

Copyright ©1999 Frank Schaffer Publications, Inc.
23740 Hawthorne Blvd.
Torrance, CA 90505

TABLE OF CONTENTS

Introduction 2
Pictographs (Analyzing data and interpreting a pictograph) 3
More Pictographs (Solving problems using a pictograph) 4
Making a Pictograph (Using data to make a pictograph) 5
Bar Graphs (Analyzing data and interpreting a bar graph) 6
Big Number Bar Graphs (Solving problems using a bar graph) 7
Decimal Amounts (Solving problems using a bar graph) 8
Making a Bar Graph (Using data to make a bar graph) 9
Double Bar Graphs (Analyzing data and interpreting a double bar graph) 10
More Double Bar Graphs (Solving problems using a double bar graph) 11
Comparing TV Habits (Using data to make a double bar graph) 12
Line Plots (Analyzing data and interpreting a line plot) 13
Making a Line Plot (Using data to make a line plot) 14
Stem and Leaf Graph
     (Analyzing data and interpreting a stem and leaf graph) 15
Math Quiz scores (Using data to make a stem and leaf graph) 16
Line Graphs (Analyzing data and interpreting a line graph) 17
More About Line Graphs (Solving problems using a line graph) 18
Making a Line Graph (Using data to make a line graph) 19
Double Line Graphs
     (Analyzing data and interpreting a double line graph) 20
Venn Diagrams (Analyzing data and interpreting a Venn diagram) 21
More Venn Diagrams (Solving problems using a Venn diagram) 22
Percents (Solving problems using a Venn diagram) 23
A Coordinate Grid (Reading a coordinate graph) 24
Ordered Pairs (Graphing ordered pairs) 25
Circle Graphs (Analyzing data and interpreting a circle graph) 26
More Circle Graphs (Solving problems using a circle graph) 27
Making a Circle Graph (Using data to make a circle graph) 28
Percents and Circle Graphs (Solving problems using a circle graph) 29
Different Graphs (Choosing the most appropriate graph) 30
Answers 31-32

Notice! Student pages may be reproduced by the classroom teacher for classroom use only, not for commercial resale. No part of this publication may be reproduced for storage in a retrieval system, or transmitted in any form or by any means—electronic, mechanical, recording, etc.—without the prior written permission of the publisher. Reproduction of these materials for an entire school or school system is strictly prohibited.

INTRODUCTION

This book has been designed to help students succeed in math. It is part of the *Math Minders* series that provides students with opportunities to practice math skills that they will use throughout their lives.

The activities in this book have been created to help students feel confident about working with graphs. The activities review skills learned in earlier grades and reinforce skills introduced and developed in fifth grade. The activities are a mixture of simple to more difficult problems. Each activity features a theme to help maintain a high level of interest for students. The activities can be used as supplemental material to reinforce any existing math program.

The skills covered in this book are pictographs, single and double bar graphs, line plots, stem and leaf plots, single and double line graphs, Venn diagrams, coordinate graphs, and circle graphs. Students will apply other math skills such as addition, subtraction, multiplication, division, fractions, percents, and other fundamental skills necessary to take the next step up the mathematical ladder. This book has been primarily designed for students to compute the answers. However, a calculator may be helpful on some of the problems.

Graphing
GRADE 5

Name_____

Pictographs
Analyzing data and interpreting a pictograph

A **pictograph** is a graph that uses pictures to show data.

This pictograph shows how long it takes the fifth grade students to get to Green Valley school.

Travel Time to School

Time	Students
5 minutes	👤👤👤👤 (4)
10 minutes	👤👤👤👤👤👤 (6)
15 minutes	👤👤👤👤👤👤👤👤👤👤 (10)
20 minutes	👤👤👤👤 (4)
25 minutes	👤👤 (2)

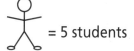 = 5 students

Use the pictograph to answer the questions.

A. What does each symbol represent? _____

B. How many students need about 10 minutes to get to school? _____

C. Which travel time amount was given by 25 students? _____

D. How many more students take about 15 minutes to get to school than students who take about 5 minutes to get to school? _____

E. Do more students take about 10 minutes or 20 minutes to get to school? _____

F. How many students' responses are shown on the pictograph? _____

G. Suppose 5 students say that it takes them 30 minutes to get to school. How many symbols should be used to show this data on the pictograph? _____

H. Write a question about the graph for a classmate to answer. _____
_____
_____

Name_____

More Pictographs
Solving problems using a pictograph

This pictograph shows the number of bottled drinks sold in one day at Quick Market.

Bottled Drinks Sold

Iced Tea	🍶 🍶 🍶 🍶 🍶 🍶 🍶 (
Soft Drinks	🍶 🍶 🍶 🍶 🍶 🍶 🍶
Water	🍶 🍶 🍶 🍶 🍶 🍶 🍶 🍶 (
Juice	🍶 🍶 🍶
Lemonade	🍶 🍶 🍶

🍶 = 10 bottles

Use the pictograph to answer the questions.

A. What does each symbol represent? _____

B. What does a half symbol represent? _____

C. Of which drink were the fewest bottles sold? _____

D. How many bottles of soft drinks were sold? _____

E. How many bottles of iced tea were sold? _____

F. How many more bottles of water were sold than iced tea? _____

G. How many symbols would be used to show 55 bottles of iced coffee sold? _____

H. Were more bottles of lemonade or juice sold? _____

I. In all, how many bottles of drinks were sold? _____

J. Write a question about the graph for a classmate to answer. _____

_____

_____

Name_____

Making a Pictograph
Using data to make a pictograph

The chart at the right shows the favorite kinds of TV shows of the fifth grade students at Pine Woods school. Use the chart to complete the pictograph below.

TV Show	Number of Students
Cartoon	|||| |||| |||| |||
Comedy	|||| |||| |||| |||| ||||
Adventure	|||| |||| ||||
News	||||	
Other	|||| ||||

Favorite Kinds of TV Shows

Cartoon	
Comedy	
Adventure	
News	
Other	

Each ☐ stands for 3 people.

Use the pictograph to answer the questions.

A. How many students voted for cartoons as their favorite kind of TV show? _____

B. Which type of show was chosen by the greatest number of students? _____

C. How many symbols were needed to show the votes for adventure shows? _____

D. How many symbols were needed to show the votes for comedy shows? _____

E. How many students voted for a favorite type of TV show? _____

F. In all, how many students chose comedy shows or adventure shows? _____

G. Which two types of shows did more than 16 students choose? _____

H. Write a question about the graph for a classmate to answer. _____

_____

_____

Name_____

Bar Graphs
Analyzing data and interpreting a bar graph

A **bar graph** is a graph that displays information using bars of different heights.
This bar graph shows the favorite team sports of the kids in the after school sports club.

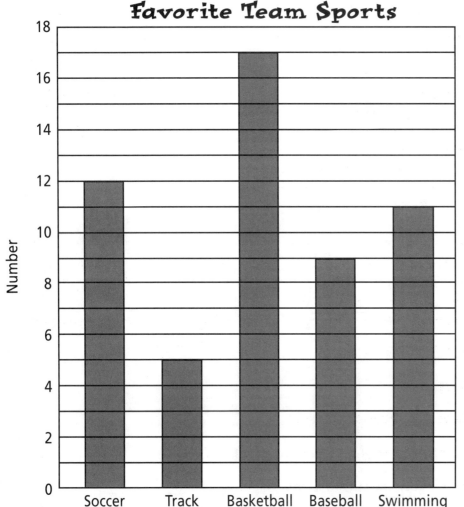

Use the bar graph to answer the questions.

A. What does each bar represent? _____

B. Which team sport got the most votes? _____

C. How many students voted for soccer as their favorite team sport? _____

D. How many students voted for baseball as their favorite team sport? _____

E. How many more students voted for basketball than for baseball? _____

F. How many more students would need to vote for swimming in order for it to have the same number of votes as soccer? _____

G. How many students' responses are shown on the bar graph? _____

© Frank Schaffer Publications, Inc. Graphing Grade 5

Name_____

Big Number Bar Graphs
Solving problems using a bar graph

This bar graph shows the pets owned by the residents of Lake View.

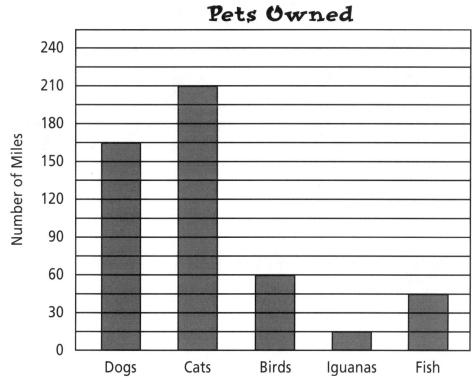

Use the bar graph to answer the questions.

A. Which pet is owned by the fewest number of people? _____

B. How many more people own cats than dogs? _____

C. How many more people own fish than iguanas? _____

D. How many pets are represented by the survey? _____

E. Are there more birds or fish? _____

F. In all, how many dogs and cats are in Lake View? _____

G. 10% of the birds are parrots. How many parrots are in Lake View? _____

H. Are there fewer iguanas or birds? _____

I. Write a question about the graph for a classmate to answer. _____

_____

Name_____

Decimal Amounts
Solving problems using a bar graph

This bar graph shows the amounts of money that can be earned doing certain jobs.

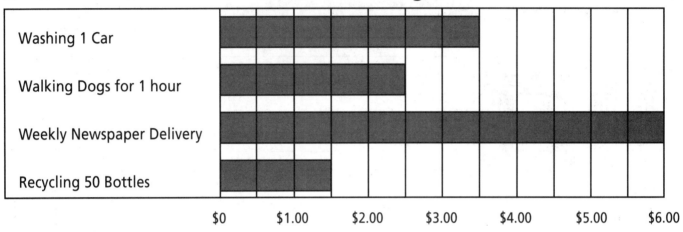

Jobs and Wages

Use the bar graph to answer the questions.

A. What is the wage for washing a car? _____

B. What earns a wage of $6.00? _____

C. What is the wage for recycling bottles? _____

D. How much more is the wage for newspaper delivery than for a car wash? _____

E. Suppose you wanted to add "mother's helper" for $2.50. What else earns the same wage?
_____

F. How much can be earned washing 5 cars? _____

G. Write a question about the graph for a classmate to answer. _____
_____

Name_____

Making a Bar Graph
Using data to make a bar graph

The chart on the right compares the approximate speeds of some land animals. Use the chart to complete the bar graph below.

Animal	Miles Per Hour
Cheetah	70
Lion	50
Elk	45
Zebra	40
Rabbit	35
Elephant	25

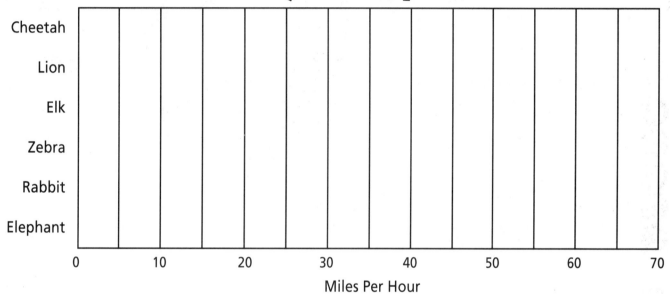

Use the bar graph to answer the questions.

A. Which animal's speed is between the speeds of the elk and the rabbit? _____

B. Do you think the table or the bar graph offers a better display of the data? Explain your answer.
_____
_____

C. Which animal has the greatest speed? _____

D. Man's speed is 27.89 miles per hour. Between which animals' speeds is man's speed?
_____

E. How many more miles per hour can a lion run than a rabbit? _____

Name_____

Double Bar Graphs
Analyzing data and interpreting a double bar graph

A **double bar graph** displays two sets of data using two sets of bars. Bars for each set usually differ in pattern, shade or color.

This double bar graph compares the number of third, fourth, and fifth grade boys and girls at Oak Meadow school.

= girls

= boys

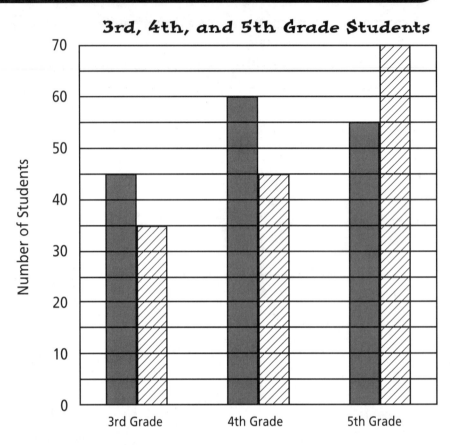

Use the double bar graph to answer the questions.

A. Which grade has the most girls? _____

B. How many students are in third grade? _____

C. Which grade has the most boys? _____

D. How many more boys are in fifth grade than in fourth grade? _____

E. Which grade has the most students? _____

F. In the fifth grade, what is the difference between number of boys and girls? _____

G. Are there more boys or girls in the third, fourth, and fifth grade classes? _____

How many more? _____

H. When is it best to use a double bar graph instead of a single bar graph? _____

I. Write a question about the graph for a classmate to answer. _____

© Frank Schaffer Publications, Inc. 10 Graphing Grade 5

Name_____

More Double Bar Graphs

Solving problems using a double bar graph

This double bar graph compares the prices of renting versus buying costumes.

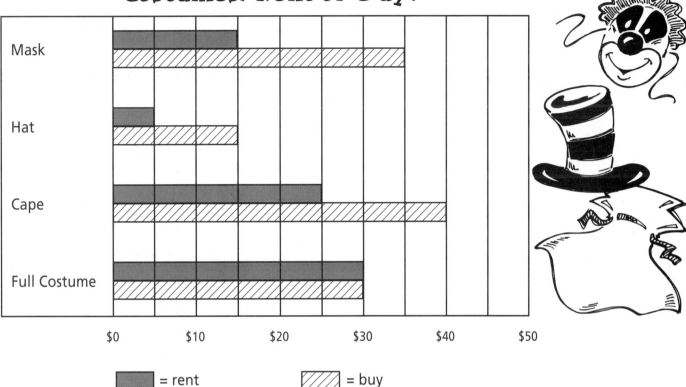

Use the double bar graph to answer the questions.

A. Which costume part costs $15 to rent? _____

B. What costs the same to rent or to buy? _____

C. Which costume part costs $15 more to buy than to rent? _____

D. Which costume part costs $15 to buy? _____

E. Which costume part costs $10 more to buy than to rent? _____

F. Which costume part costs $20 more to buy than to rent? _____

G. How much does it cost to rent a hat and a mask? _____

H. How much more does it cost to buy a cape and a hat than it costs to rent them? _____

I. Write a question about the graph for a classmate to answer. _____
_____

Name_____

Comparing TV Habits
Using data to make a double bar graph

The chart at the right shows the hours of TV watched in one day by the fifth grade and sixth grade students at Miller Middle school. Use the chart to complete the double bar graph below.

Hours Per Day	5th Grade	6th Grade
1 or less	4	12
2	7	10
3 or more	14	3

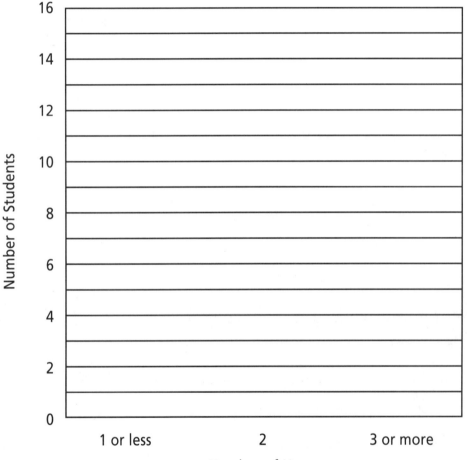

Use the double bar graph to answer the questions.

A. How many students were surveyed in 5th grade? _____

In 6th grade? _____

B. How many hours of TV do the greatest number of 5th graders watch?

_____

C. How many hours of TV does the greatest number of 6th graders watch? _____

D. Which grade watches more total hours of TV per day? _____

E. How many 6th graders watch 3 or more hours of TV per day? _____

F. Twelve 6th graders watch 1 hour or less of TV per day. 25% of them are boys. How many 6th grade boys watch 1 or less hours of TV per day?

_____

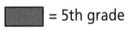 = 5th grade   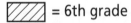 = 6th grade

Graphing Grade 5

Name_____

Line Plots
Analyzing data and interpreting a line plot

A **line plot** is a graph that shows data by using symbols that are lined up.

This line plot shows how many visits each of the students in the after school science club made to the school media center in one month.

School Media Center Visits

```
                        X
                        X
                        X              X
                        X              X
                        X              X        X
  X                     X              X        X
  X           X         X     X        X        X
  X           X         X     X        X        X
  X           X         X     X        X        X

  1     2     3         4     5        6        7
                     Number of Visits
```

Use the line plot to answer the questions.

A. What was the most common number of visits made by students this month? _____

B. How many students visited the media center 6 times during the month? _____

C. In all, how many students visited the Media Center 3, 4, or 5 times during the month? _____

D. How many students' responses are shown on the line plot? _____

E. Are there any gaps in the line plot? _____

If so, what does it mean? _____

F. Write a fraction to show how many students visited the media center 4 times in the month. (Write it in lowest terms.) _____

G. How many times did 25% of the students visit the media center during the month? _____

H. Write a question about the graph for a classmate to answer. _____
_____

© Frank Schaffer Publications, Inc.   Graphing Grade 5

Making a Line Plot

Using data to make a line plot

Brad asked 24 of his friends how many brothers or sisters they have. The table at the right shows their responses. The number in each box is the number given by each friend. Use the table to complete the line plot. Use an **X** to represent each person.

Number of Siblings							
0	3	2	1	0	0	1	2
1	2	3	1	0	1	1	2
0	1	2	1	5	1	3	0

Number of Siblings

0   1   2   3   4   5

Use the line plot to answer the questions.

A. How many students do not have any brothers or sisters? _____

B. How many students have one sibling? _____

C. How many students have more than three siblings? _____

D. How many more students have one sibling than those who have two siblings? _____

E. How many students' responses are shown in the line plot? _____

F. Write a fraction to show how many of the 24 students have 3 siblings. (Write it in lowest terms.) _____

G. Add an **X** to the line plot to show the number of siblings you have. Write a sentence to tell how the line plot changed. _____

Name_____

Stem and Leaf Graph
Analyzing data and interpreting a stem and leaf graph

A **stem and leaf graph** is an arrangement of data with numbers separated into tens and ones, with tens lined up in a stem formation and ones branching off to the side like leaves. The table at the right shows the heights of the boys in Mr. Wilson's fifth grade class.

Heights of Fifth Grade Boys in Inches
44, 48, 52, 53, 55, 56, 56, 57, 58, 59, 60, 62

This stem and leaf graph shows the same information as in the table above.

Heights of Fifth Grade Boys in Inches

stem	leaves
4	4, 8
5	2, 3, 5, 6, 6, 7, 8, 9
6	0, 2

Key: 5|2 means 52 inches.

Use the stem and leaf graph to answer the questions.

A. What numbers are the "stems" of the stem and leaf graph? _____

B. What numbers are the "leaves" of the "4" stem? _____

C. Which stem has the most leaves? _____

D. How many fifth grade boys' heights are shown on the stem and leaf graph? _____

E. Which height describes more than one boy? _____

F. Suppose a new fifth grade boy is 49 inches tall. Tell how his height can be added to the stem and leaf graph. _____

G. What do you know about the heights of the fifth grade boys surveyed by reading the stem and leaf graph? _____

H. Write a question about the graph for a classmate to answer. _____

Name_____

Math Quiz Scores
Using data to make a stem and leaf graph

The table at the right shows the math quiz scores of the students in the math club. Use the table to complete the stem and leaf graph.

Math Quiz Scores
78, 78, 80, 80, 82, 84, 88, 89, 90, 90, 91, 91, 92, 92, 92, 93, 94, 95, 95, 96, 96, 97, 98, 100

Math Quiz Scores

stem	leaves
___	_____
___	_____
___	_____
10	0

Key: 8|5 means a score of 85.

Use the stem and leaf graph to answer the questions.

A. What numbers are the "stems" of the stem and leaf graph? _____

B. What numbers are the "leaves" of the "8" stem? _____

C. Which stem has the most leaves? _____

D. Which stem has the fewest leaves? _____

E. Which score did three students receive on the quiz? _____

F. How many students' scores are shown on the stem and leaf graph? _____

G. Did any students receive a score of 81? _____

H. How many students received a perfect score? _____

I. Add a score of 99 to the stem and leaf graph. Where did you plot the score? _____

J. What do you know about the scores of the math quiz by reading the stem and leaf graph? _____

K. Write a question about the graph for a classmate to answer. _____

© Frank Schaffer Publications, Inc.   Graphing Grade 5

Name_____

Line Graphs
Analyzing data and interpreting a line graph

A **line graph** shows data that changes over time.

This line graph shows the number of miles Marci ran each week during an eight-week period.

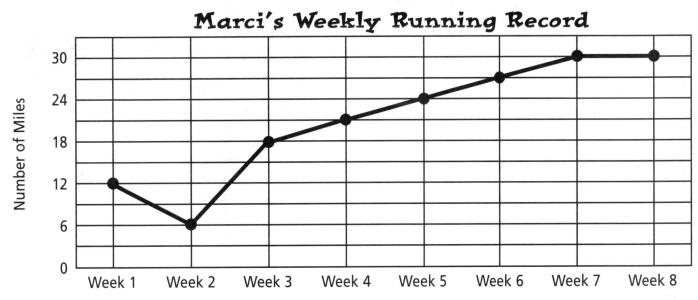

Use the line graph to answer the questions.

A. How many miles did Marci run during week 2? _____

B. How many miles did Marci run during week 5? _____

C. During which week did the number of miles decrease? _____

D. During which week did the number of miles increase the most? _____

E. How many miles did Marci run in all during the eight weeks? _____

F. How many more miles did Marci run during week 5 than during week 2? _____

G. What was the average number of miles Marci ran per week? _____

H. Predict the number of miles Marci will run in week 9. Give an explanation for your answer. _____
_____

I. Write a question about the graph for a classmate to answer. _____
_____

Name_____

More About Line Graphs
Solving problems using a line graph

This line graph shows the number of video sales of a popular movie called *Smiley* during a seven-month period.

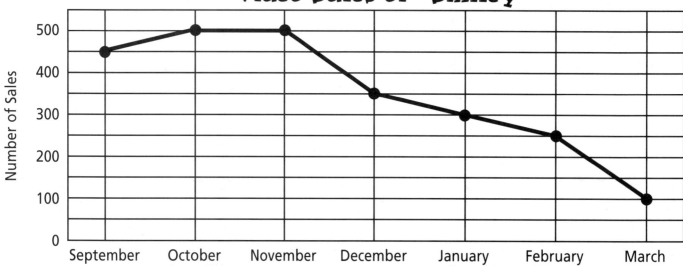

Use the line graph to answer the questions.

A. Which month did the number of sales show an increase? _____

B. During which two months did the number of sales remain the same? _____

C. How many sales were made in December? _____

D. How many sales were made in March? _____

E. How many more sales were made in November than in January? _____

F. What was the average number of videos sold each month? _____

G. Describe the trend that the line graph shows. _____
_____

H. This video store can order 500 more copies of Smiley at a sale price. Do you think they should order them? Justify your answer. _____
_____

I. Write a question about the graph for a classmate to answer. _____
_____

Name_____

Making a Line Graph
Using data to make a line graph

The table at the right shows the number of sport utility vehicle sales at Sporty Car Sales during a six-month period. Use the table to complete the line graph below.

Sport Utility Vehicle Sales	
October	12
November	24
December	30
January	36
February	24
March	6

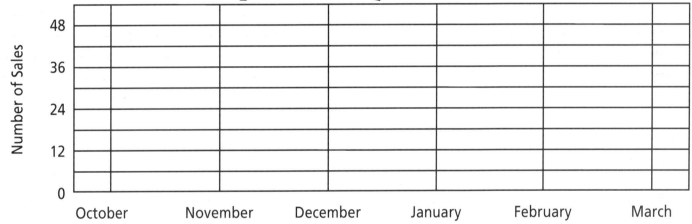

Use the line graph to answer the questions.

A. Which month did the number of sales increase the most? _____

B. Which month did the number of sales decrease the most? _____

C. During which two months were the sales the same? _____

D. How many sport utility vehicles were sold in December? _____

E. In all, how many cars were sold during the six-month period? _____

F. What was the average number of sport utility vehicles sold each month? _____

G. How many more sport utility vehicles were sold in January than in October? _____

H. Write a question about the graph for a classmate to answer. _____

Name_____

Line Graphs
Analyzing data and interpreting a double line graph

A **double line graph** compares two sets of data over the same period of time.
This double line graph shows the number of boys and girls who participated in sports between the years of 1991 and 1998.

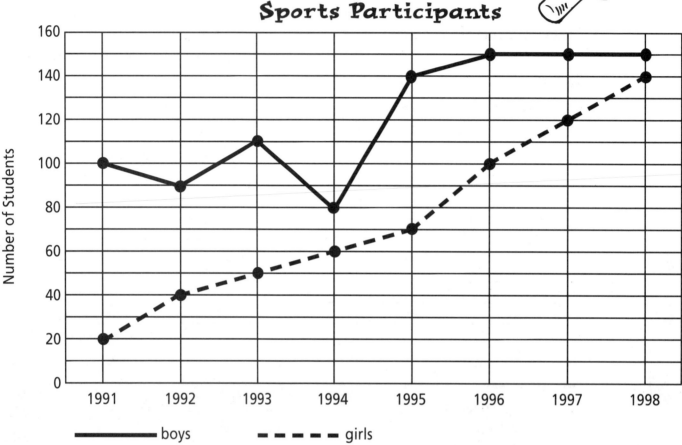

Use the double line graph to answer the questions.

A. What year had the smallest number of boys in sports? _____

B. What year had the greatest number of girls in sports? _____

C. Which year did the number of boys in sports decrease the most? _____

D. Which year did the number of girls in sports increase the most? _____

E. How many more girls participated in sports in 1998 than in 1991? _____

F. Predict what will happen to the number of boys and girls in sports in the next two years. Look at the trend for boys and girls to help you. _____

Name_____

Venn Diagrams
Analyzing data and interpreting a Venn diagram

A **Venn diagram** is a drawing that uses geometric shapes to show relationships between sets of data and how the information may overlap.

This Venn diagram shows the number of students who take part in certain after school activities.

After School Activities

Sports — Music

- Sports only: 10
- Sports ∩ Music: 15
- Music only: 7
- Sports ∩ Scouts: 10
- All three: 6
- Music ∩ Scouts: 7
- Scouts only: 3

Scouts

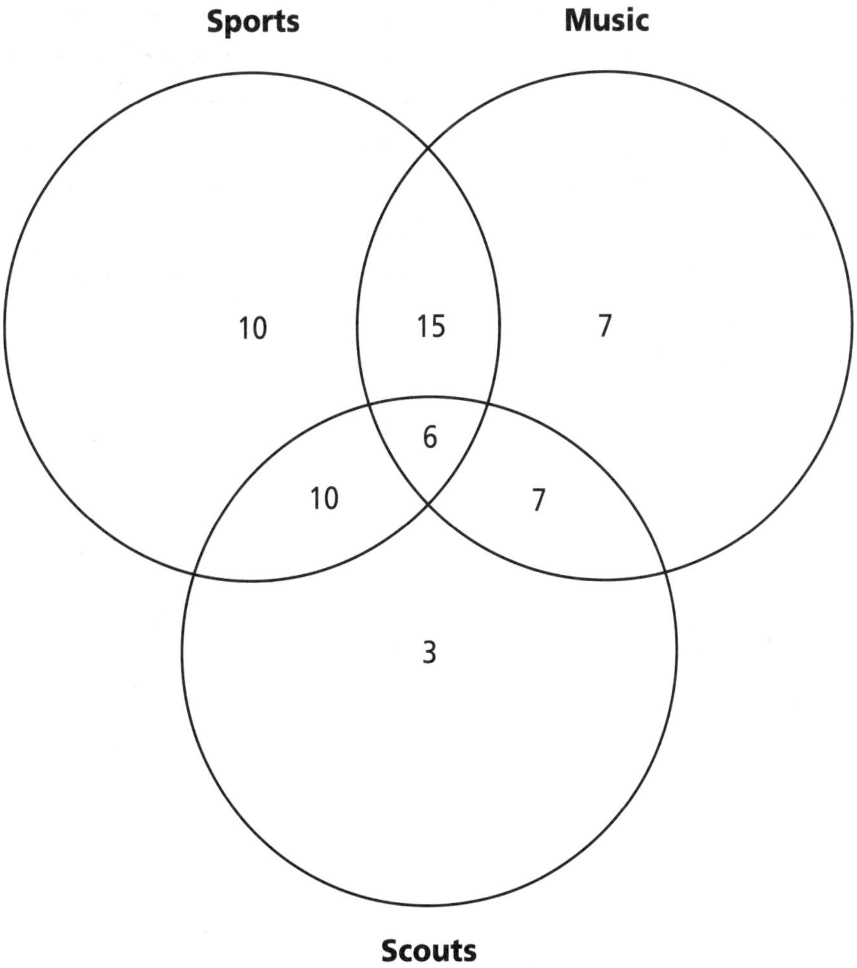

Use the Venn diagram to answer the questions.

A. How many students play sports? _____

B. How many students play musical instruments? _____

C. How many students belong to scouts? _____

D. How many students participate in sports and also play musical instruments? _____

E. How many students are in scouts and also play sports? _____

F. How many students only play musical instruments? _____

G. How many students participate in after school activities? _____

H. How many students participate in all three activities? _____

I. 60% of the students who only play sports are boys. How many boys only play sports? _____

J. Write a question about the Venn diagram for a classmate to answer. _____

_____

_____

Name_____

More Venn Diagrams
Solving problems using a Venn diagram

This Venn diagram shows the number of siblings and pets a group of fifth grade students have.

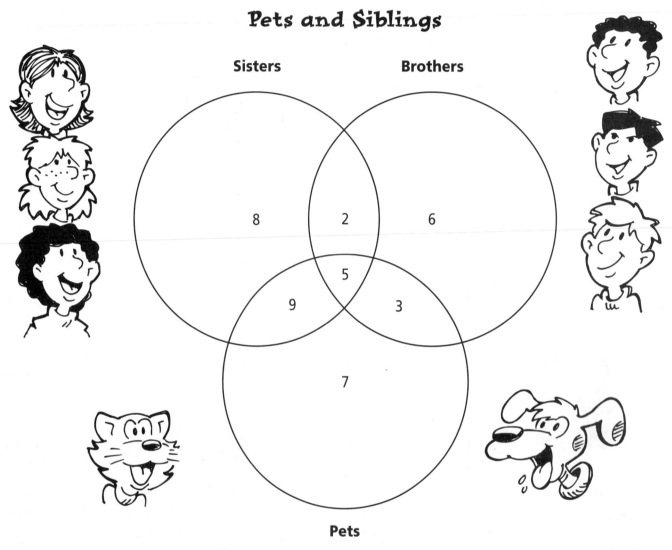

Use the Venn diagram to answer the questions.

A. How many students have sisters? _____

B. How many students have brothers? _____

C. How many students have pets? _____

D. How many students have both sisters and brothers? _____

E. How many students have both sisters and pets? _____

F. How many students have sisters, but no brothers or pets? _____

G. How many students have sisters, brothers, and pets? _____

Name_____

Percents
Solving problems using a Venn diagram

Venn diagrams can be used to show percentages. This Venn diagram shows the clubs joined by 100 students at a neighborhood school.

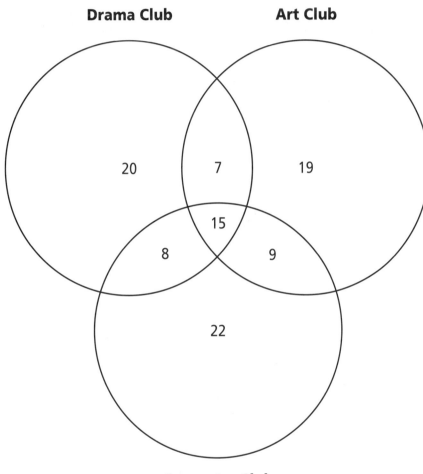

Club Members

Drama Club — 20
Drama ∩ Art — 7
Art Club — 19
Drama ∩ Computer — 8
All three — 15
Art ∩ Computer — 9
Computer Club — 22

Use the Venn diagram to answer the questions.

A. What percentage of the students are in the art club? _____

B. What percentage of the students are in the drama club? _____

C. What percentage of the students are in the computer club? _____

D. What percentage of the students are in all three clubs? _____

E. What percentage of the students are in only one club? _____

F. What percentage of the students are in exactly two clubs? _____

G. What percentage of the students are in both the art club and the computer club? _____

H. Add the percentages for questions A, B, and C. Do they combine to make 100%? _____

How can you explain it? _____

Name_____

A Coordinate Grid
Reading a coordinate graph

A **coordinate graph** displays information on a grid. Numbered pairs give a point's location on a grid. The first number in each ordered pair tells how many numbers across. The second number tells how many spaces up.

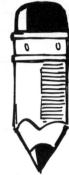

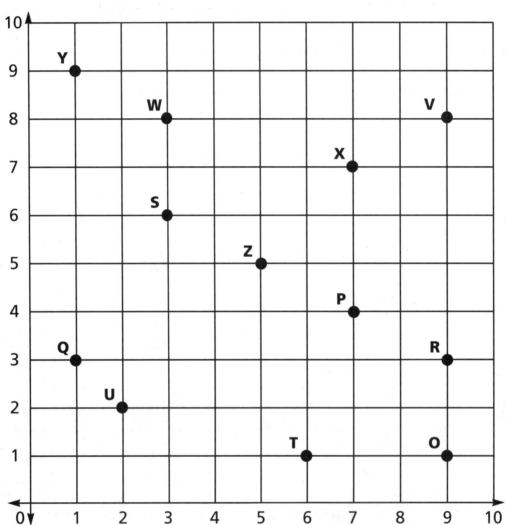

Circle the correct ordered pair for each point.

A.	Z	(1, 6)	(5, 5)	(1, 1)		**G.**	T	(1, 6)	(5, 4)	(6, 1)
B.	Y	(1, 9)	(9, 1)	(7, 4)		**H.**	S	(6, 3)	(6, 2)	(3, 6)
C.	X	(3, 3)	(2, 1)	(7, 7)		**I.**	R	(7, 2)	(9, 3)	(7, 6)
D.	W	(8, 3)	(3, 8)	(3, 1)		**J.**	Q	(5, 4)	(3, 1)	(1, 3)
E.	V	(9, 8)	(2, 6)	(6, 2)		**K.**	P	(2, 7)	(7, 4)	(4, 7)
F.	U	(5, 6)	(2, 2)	(2, 1)		**L.**	O	(9, 1)	(9, 9)	(1, 1)

© Frank Schaffer Publications, Inc. Graphing Grade 5

Name_____

Ordered Pairs
Graphing ordered pairs

Use the ordered pairs to locate each point on the grid. Label the points in order from A to N. Remember, first count across, then up.

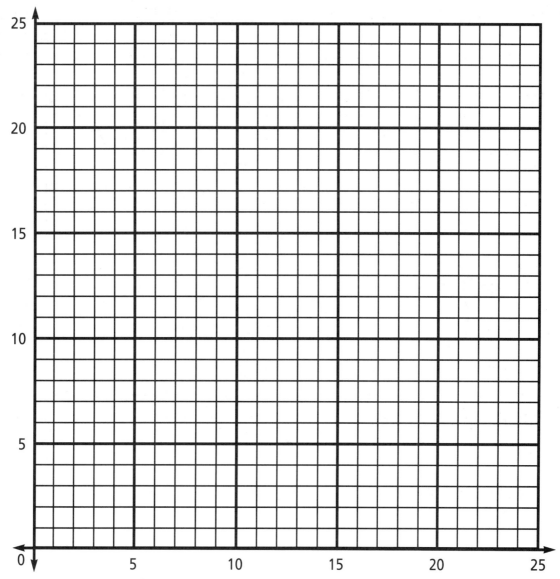

- **A.** (10, 2)
- **B.** (2, 7)
- **C.** (10, 20)
- **D.** (20, 20)
- **E.** (22, 10)
- **F.** (10, 15)
- **G.** (18, 9)
- **H.** (5, 5)
- **I.** (23, 2)
- **J.** (17, 4)
- **K.** (22, 17)
- **L.** (12, 13)
- **M.** (3, 23)
- **N.** (15, 2)

Name_____

Circle Graphs
Analyzing data and interpreting a circle graph

A **circle graph** is a graph in the form of a circle that is divided into sections showing how the whole is broken into parts.

This circle graph shows how Brad spends his monthly allowance.

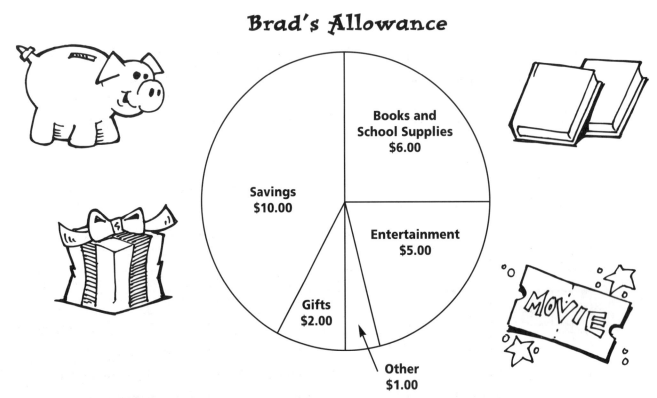

Use the circle graph to answer the questions.

A. For what is the largest portion of the allowance used? _____

B. How much does Brad get for a monthly allowance? _____

C. For what is $\frac{1}{4}$ of the allowance used? _____

D. What portion of the allowance is set aside for other uses? _____

E. What percentage of the allowance is spent on books and school supplies? _____

F. How much is spent of gifts? _____

G. How much more is spent on entertainment than for other uses? _____

H. How much does Brad save during a six-month period? _____

I. Write a question about the graph for a classmate to answer. _____

_____

© Frank Schaffer Publications, Inc.   Graphing Grade 5

Name_____

More Circle Graphs
Solving problems using a circle graph

The circle graph shows the favorite zoo animals of 240 students.

Favorite Zoo Animals

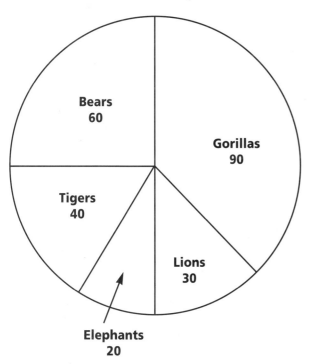

Use the circle graph to answer the questions.

A. Which animal was chosen by the greatest number of students? _____

B. Which animal was chosen by the fewest students? _____

C. How many students chose bears as their favorite zoo animal? _____

D. How many students chose lions as their favorite zoo animal? _____

E. How many students chose gorillas as their favorite zoo animal? _____

F. How many more students chose tigers than lions as their favorite zoo animal? _____

G. How many students did not choose gorillas as their favorite zoo animal? _____

H. Which animal was chosen by 25% of the students? _____

I. Which animal was chosen by three times as many students as the elephants? _____

J. Which animal was chosen by half as many students as the bears? _____

K. Write a question about the graph for a classmate to answer. _____

_____

_____

_____

Name_____

Making a Circle Graph
Using data to make a circle graph

The chart below shows the portions of income received from one year of fundraising at Redgate School. A total of $1,000 was raised. Use the data to complete the circle graph below.

First choose a color to represent each category and color the key. Then color each section of the circle to match the information on the chart.

Fundraiser Sales

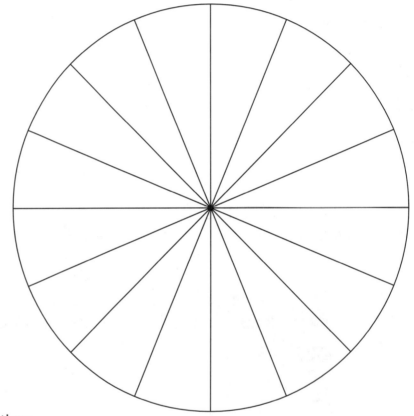

Fundraiser Sales	
Wrapping Paper	1/8
Greeting Cards	1/16
Candy	1/4
Pizza	5/16
Books	1/4

☐ = wrapping paper
☐ = greeting cards
☐ = candy
☐ = pizza
☐ = books

Use the circle graph to answer the questions.

A. Sales of which item raised the largest portion of money? _____

B. Sales of which item raised the smallest portion of money? _____

C. Sales of which two items raised the same amount of money? _____

D. Sales of which item raised less than $100? _____

E. Sales of which items raised between $200 and $300? _____

F. Write a question about the graph for a classmate to answer. _____

_____

© Frank Schaffer Publications, Inc.

Graphing Grade 5

Name_____

Percents and Circle Graphs
Solving problems using a circle graph

Circle graphs can be used to show percentages of a group of items. The whole circle represents 100%, or the whole amount. The chart at the right shows the percentages of sales for each type of CD during one month at Music Mart. Use the data in the table to complete the circle graph below.

First choose a color to represent each category and color the key. Then color each section of the circle to match the information on the chart.

CD Sales in December	
Music	40%
Books on CD	10%
Education	20%
Business	20%
Other	10%

☐ = music
☐ = books on CD
☐ = education
☐ = business
☐ = other

Use the circle graph to answer the questions.

A. Sales of which type of CD represented the largest percentage? _____

B. If total CD sales were $5,000, how much money was sold in music CDs? _____

C. If total CD sales were $10,000, how much money was sold in education CDs? _____

D. If total CD sales were $3,000, how much money was sold in business CDs? _____

E. Write a question about the graph for a classmate to answer. _____
_____
_____
_____
_____

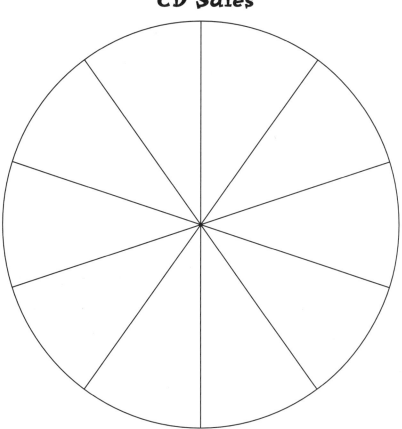

CD Sales

© Frank Schaffer Publications, Inc.

Graphing Grade 5

Name_____

Different Graphs
Choosing the most appropriate graph

There are many kinds of graphs. When you are trying to decide which type of graph to use, think about what kind of display would give the clearest information.

Write which type of graph you would choose to show the following information.

A. You want to show results of a survey of 10,000 people.

_____

B. You surveyed 50 students to find out what kind of pets they have.

_____

C. You want to show how many people visited the library each day for one week.

_____

D. You want to show the average temperature in a particular city.

_____

E. You want to graph the heights of your five best friends.

_____

F. You want to graph your neighbors' houses to show where they are in location to your home.

_____

G. You want to show how much money you earned mowing the lawn each week for a six-week period.

_____

H. You helped your mom plant 100 flower bulbs. You want to show how many of each type was planted.

_____

I. You want to show how many tacos were sold last year at Tommy's Taco Hut.

_____

bar graph

line graph

circle graph

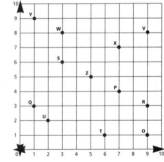
coordinate graph

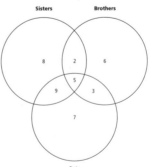
Venn diagram

© Frank Schaffer Publications, Inc. Graphing Grade 5

ANSWERS

Page 3
A. 5 students
B. 30 students
C. 20 minutes
D. 25 students
E. 10 minutes
F. 130 students
G. 1 symbol
H. Questions will vary.

Page 4
A. 10 bottles
B. 5 bottles
C. lemonade
D. 80
E. 75
F. 20
G. 5 1/2
H. juice
I. 320
J. Questions will vary.

Page 5
Graph should be completed as shown:

Favorite Kinds of TV Shows	
Cartoon	☐☐☐☐☐☐
Comedy	☐☐☐☐☐☐☐☐
Adventure	☐☐☐☐
News	☐☐
Other	☐☐☐

A. 18
B. comedy
C. 5
D. 8
E. 72
F. 39
G. cartoon and comedy
H. Questions will vary.

Page 6
A. votes for a favorite team sport
B. basketball
C. 12
D. 9
E. 8
F. 1
G. 54

Page 7
A. iguanas
B. 45
C. 30
D. 495
E. birds
F. 375
G. 6
H. iguanas
I. Questions will vary.

Page 8
A. $3.50
B. Weekly newspaper delivery
C. $1.50
D. $2.50
E. Walking dogs for 1 hour
F. $17.50
G. Questions will vary.

Page 9
Graph should be completed as shown:

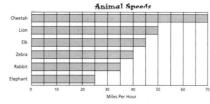

A. zebra
B. Answers vary. Possible answer: The bar graph shows the data more clearly because you can quickly see all the animals' speeds relative to each other.
C. cheetah
D. rabbit and elephant
E. 15 miles

Page 10
A. 4th grade
B. 80
C. 5th grade
D. 25
E. 5th grade
F. 15
G. girls; 10
H. when there are two sets of data to compare
I. Questions will vary.

Page 11
A. mask
B. full costume
C. cape
D. hat
E. hat
F. mask
G. $20
H. $25
I. Questions will vary.

Page 12
Graph should be completed as shown:

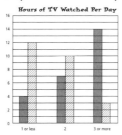

A. 25; 25
B. 3 or more
C. 1 or less
D. 5th grade students
E. 3
F. 3

Page 13
A. 4
B. 6
C. 13
D. 28
E. yes; No students visited the media center 2 times during the month.
F. 1/4
G. 4
H. Questions will vary.

Page 14
Graph should be completed as shown:

A. 6
B. 9
C. 1
D. 4
E. 24
F. 1/8
G. 0
H. An additional X should be added to the graph in the appropriate place. Sentences will vary.

Page 15
A. 4, 5, and 6
B. 4 and 8
C. 5
D. 12
E. 56 inches
F. Add a 9 "leaf" to the "4" stem.
G. Possible answer: Heights of most 5th grade boys are between 50 and 59 inches.
H. Questions will vary.

Math Quiz Scores	
Stem	**Leaves**
7	8, 8
8	0, 0, 2, 4, 8, 9
9	0, 0, 1, 1, 2, 2, 2, 3, 4, 5, 5, 6, 6, 7, 8
10	0

Page 16
A. 7, 8, 9, and 10
B. 0, 0, 2, 4, 8, and 9
C. 9
D. 10

ANSWERS

E. 92
F. 24
G. no
H. 1
I. A "9" leaf is added to the "9" stem.
J. Possible answer: Most scores fell between 90 and 99.
K. Questions will vary.

Page 17
A. 6
B. 24
C. Week 2
D. Week 3
E. 168
F. 18
G. 21
H. Accept reasonable answers.
I. Questions will vary.

Page 18
A. October
B. October and November
C. 350
D. 100
E. 200
F. 350
G. Possible answer: The line graph increases between September and October, stays the same through November, and then decreases every month from December through March.
H. Accept reasonable answers.
I. Questions will vary.

Page 19
Graph should be completed as shown:

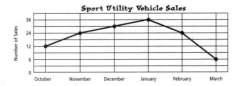

A. November
B. March
C. November and February
D. 30
E. 132
F. 22
G. 24
H. Questions will vary.

Page 20
A. 1994
B. 1998
C. 1994
D. 1996
E. 120
F. Accept reasonable answers.

Page 21
A. 41
B. 35
C. 26
D. 21
E. 16
F. 7
G. 58
H. 6
I. 6
J. Questions will vary.

Page 22
A. 24
B. 16
C. 24
D. 7
E. 14
F. 8
G. 5

Page 23
A. 50%
B. 50%
C. 54%
D. 15%
E. 61%
F. 24%
G. 24%
H. no; Possible answer: A, B, and C combine to make more than 100% because students who joined more than one club are counted more than once.

Page 24
A. (5, 5)
B. (1, 9)
C. (7, 7)
D. (3, 8)
E. (9, 8)
F. (2, 2)
G. (6, 1)
H. (3, 6)
I. (9, 3)
J. (1, 3)
K. (7, 4)
L. (9, 1)

Page 25
Grid should be completed as shown:

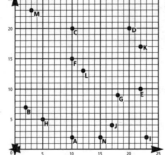

Page 26
A. savings
B. $24.00
C. books and school supplies
D. $1.00
E. 25%
F. $2.00
G. $4.00
H. $60.00
I. Questions will vary.

Page 27
A. gorillas
B. elephants
C. 60
D. 30
E. 90
F. 10
G. 150
H. bears
I. bears
J. lions
K. Questions will vary.

Page 28
Graph should be colored as follows:
  wrapping paper—2 sections colored
  greeting cards—1 section colored
  candy—4 sections colored
  pizza—5 sections colored
  books—4 sections colored
A. pizza
B. greeting cards
C. candy and books
D. greeting cards
E. candy and books
F. Questions will vary.

Page 29
Graph should be colored as follows:
  music—4 sections colored
  books on CD—1 section colored
  education—2 sections colored
  business—2 sections colored
  other—1 section colored
A. music
B. $2,000
C. $2,000
D. $600
E. Questions will vary.

Page 30
Accept all reasonable answers.